The 20 Key Technologies of Industry 4.0 and Smart Factories

The Road to the Digital Factory of the Future

TABLE OF CONTENTS

1. INDUSTRY 4.0. OVERVIEW

In the next 5 years the automotive factories will change more than in the last 50 years. This 4th Industrial Revolution (called **Industry 4.0** in Europe and **Smart Industry** in the US) will be driven by the automotive sector, change radically the industrial model and soon come to other sectors.

Companies realize that this new industrial revolution is here to stay, they will survive and win the battle against competitors in low cost countries. Companies that believe the Industry 4.0 is a fad, are likely to disappear in a few years, even if they believe that they are well established today. In these moments it is more important **to use the compass** than to have the clock set at the correct hour.

Large companies are aware that the manufacturing industry is facing an unprecedented challenge and are devoting enormous resources to promote technological change, but for **SMEs** which are the main generators of employment (SMEs account for two thirds of all jobs worldwide) the effort of this transition can be overwhelming, so it is vital for them **to find the right solution and to get it right the first time**. Therefore they need access to all possible information technology, something that has come to be called democratization of knowledge and will have a multiplier effect on economic recovery, because when human knowledge is shared, it is multiplied.

In Germany, as an example, they have had the strategic vision to promote the conversion of the

Mittelstand (small and medium German companies) to industry 4.0 as a key objective of the *Industrie 4.0* plan, and this has notably helped the recovery of employment, going from an unemployment rate of 8% in 2010 to 4% today.

This new industrial model implies a great change in the paradigm of manufacturing and launches us into a seamless connection of the digital and real worlds. There are numerous tools and technologies (KETs) in this new model of Industry 4.0, which   will spread the use of **IoT** in industrial processes, and further the integration of **Cyber - Physical Systems** (CPS) horizontally (in the value chain) and vertically (from sensorization of process parameters to real time decision making).

In this model Data- Centric industry (strongly based on data)   particularly   relevant   technologies   are   **Big Data** and **Advanced Analytics**, which   detect inefficiencies, predict failures, optimize and simulate industrial processes, etc., along with technologies to develop intelligent processes (automated, flexible and self - learning).

The aim of this book is to give the reader an approximation to these new technological concepts (which separately do not seem to have an obvious connection, but do when they are used together in an industrial application context), not from a purely theoretical point but using a large  contribution of examples and practical cases to facilitate internal reflection on *"What path do we want our companies to follow and what would the next steps to take be?*

It is possible that some of the solutions that are discussed in this book could be difficult to implement because of the size of our company, but simple observation and reflection about what they are doing now in large companies provide us with fundamental knowledge to trace our own route. This work has a dual strategic vision: a) to raise awareness and practically demonstrate the potential of industry 4.0; and b) to make the leaders and managers of **SMEs** reflect upon their products, processes and business models, under the focus of innovation of Industry 4.0.

This book is principally enriched with the vision of the automotive sector but its content is applicable to all industry in general and aims to provide technological knowledge that will help us establish the roadmap needed to change our business environment and deal with the future with a guarantee of total success.

"When we thought we had all the answers, suddenly, they changed all the questions."

Mario Benedetti.

2. THE FACTORY OF THE FUTURE

The Factory of the Future is a safe, healthy space in which people and collaborative robots coexist, with continuous exploitation of all the data of the processes and a predictive approach based on Artificial Intelligence technology, in which people at all levels are proactive and motivated by the use of technology, because it gives them autonomy, mastery (the desire to be more and more capable) and allows them to achieve great impact. And that motivation generated by technology in people makes them learn more easily, overcome difficulties and achieve the impossible.

In this new factory model *total manufacturing costs* will be reduced significantly and other benefits will also be achieved, such as greater flexibility, quality, speed and communication at all levels.

Vision of the Factory of the Future for PSA

"I'm interested in the future because it's the place where I'm going to spend the rest of my life."
Woody Allen

WHY DO WE HAVE TO GO FOR THE FACTORY OF THE FUTURE?

In a report on Industry 4.0, *Tecnalia* lists the six socioeconomic trends for which we have to commit for the Factory of the future at this time:

I. The **availability of new technologies** that will allow us to reach goals unthinkable until now;

II. The **scarcity of natural resources**, so efficiency is key to deal with the new situation;

III. The need to adopt measures that favor **environmental sustainability;**

IV. Another key factor is **the aging population and late retirement,** which leads to a physical work reduction;

V. The **specialization of the staff in specific areas** of the factory itself will be more and more necessary, depending on its activity, which continually requires adapting the level of qualification of the workers and therefore forces us to rethink the traditional model;

VI. And, finally, **customization** is already a reality (demand for products adapted to the specific needs of each client), which clashes with traditional serial production where the axis was the production chain instead of the product

Siemens plant in Amberg (Germany)

WHAT WILL THE FACTORY OF THE FUTURE BE LIKE?

Automated.- To gain speed in repetitive processes, with collaborative robots (cobots) performing tasks of lower added value and easily changing from one task to another.

Digital (connected) .- Incorporating electronics to capture data in a massive way and manage production processes in real time.

Intelligent.- Interpreting the data of the processes and facilitating the decision making in advance, promoting the Continuous Improvement and Innovation.

Flexible.- In design (involving customers), in manufacturing and in logistics, to adapt immediately to changes in demand.

Sustainable.- With a rational and responsible use of resources and energy use.

Human.- All the foregoing will only be possible if we properly train our employees and activate their talent. People will continue to be the center of activity and will make a difference in this new scenario.

Sources: Sisteplant; PSA; Tecnalia; Siemens [2017]

"Every once in a while, a new technology, an old problem, and a big idea turn into an innovation." *Dean Kamen*

3. TRANSITION TO THE FACTORY OF THE FUTURE

3.1 AREAS OF IMPROVEMENT AND LEVERS FOR CHANGE

The necessary transition to Industry 4.0 has been defined in a clever way by CEAGA * e INOVA with the *Digital Transformation Wheel:*

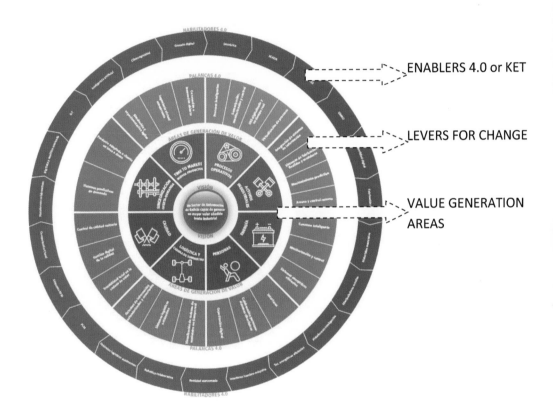

ENABLERS 4.0 or KET

LEVERS FOR CHANGE

VALUE GENERATION AREAS

* CEAGA is the Automotive Cluster of Galicia, which in 2013 won the Gold recognition Label accrediting it as one of the 22 best clusters in the European Union (EU), becoming the only cluster in the automotive industry to achieve this mark of excellence.

This wheel shows *Value Generation Areas* and *Levers* which we need to move or activate to bring about change in our organization to achieve Industry 4.0:

OPERATIONAL PROCESSES

(processes for the manufacture of products and services):

Levers for change:

intelligent processes;
sensoring, monitoring and control;
digital OEE and in real time;
virtualization.

INDUSTRIAL ASSETS

(Machinery and equipment involved in business processes):

Levers for change:

integration of information systems;
flexible systems, modular manufacturing;
predictive maintenance; access and remote control.

ENERGY

(Energy consumption and generation methods):

Levers for change:

intelligent consumption;
monitoring and control;
energy efficient systems.

PEOPLE

(Human resources involved in the management and operation of companies):

Levers for change:

interfaces;
cyber - physical collaboration between *persons and systems;*
digital training.

LOGISTICS AND SUPPLY CHAIN

(Processes and assets used to transport supplies from suppliers, moving parts inside the plant and the distribution of products to customers):

Levers for change:

standardized systems and connected information;
autonomous logistics systems;
in real time virtualized supply chain.

QUALITY

(Methods used to improve and control the quality of products and services):

Levers for change:

unit quality control;
digital quality management;
full traceability in the value chain.

SYNCHRONIZATION OFFER-DEMAND

(Ability of companies to adapt their products and services to customer requirements):

Levers for change:

product tailored to the customer based data;
predictive demand systems.

TIME-TO-MARKET OF NEW PRODUCTS

(Ability of companies to bring to market new products and services and get feedback from customers):

Levers for change:

co - creation and open innovation;
concurrent virtual engineering;
agile simulation and experimentation.

3.2   STRATEGY NEEDED FOR BUSINESSES

In order to successfully define the strategy to be followed, it is necessary to previously identify our TECHNOLOGICAL MATURITY INDEX. To do this in this book a simple test is provided in chapter 5 that will give us a score in each of the 8 areas for improvement or value generation we have defined. After completing this self - assessment process and developing a *radar chart* with the result, we will have identified the weakest areas, and then it will be time to reflect on what priorities will set to achieve maximum competitiveness.

The project, with a clear definition of where we want to go, should be ambitious but applicable, to avoid the risk of running into dead ends.

The most important milestone is the **definition of OBJECTIVES:**

Taking the *test maturity* itself, we will establish the objectives we want to achieve in the short, medium and long term in each of the 24 sub - sections. To do this we must reflect on what actions will make us become more competitive and which ones, although important, we can leave for a second or third phase. In this reflection we can ask ourselves questions like:

Are there business objectives or customer demands that require 4.0 measures?

For example, the need for higher quality, better time to market, more innovation, etc.

What operations can we focus on from a 4.0 perspective?

For example, cost reduction   (derived from poor quality, frequent breakdowns, an ineffective employment of energy, etc.)

Parallel to the process of defining the new strategy, it is fundamental for SMEs to consider the possibility of combining efforts, resources and talent. Globalization has caused a paradigm shift in the relationship between many SMEs, which have moved from competition to collaboration. Instead of disputing a local small market among themselves, it is more efficient to assume joint projects that generate greater competitiveness, which is essential for survival in a global market, competing with *low* industrial *cost* areas anywhere in the world.

To accomplish this, it is very interesting to **create clusters,** business associations that will allow us:

- To reduce costs, thanks to the possibility of sharing resources, capabilities and knowledge.
- Access common markets, mainly international or those difficult to access.
- To share activities and promotional costs.
- To increase sales. Thanks to the union, they can cover larger projects and access some that would not otherwise be possible.
- To act jointly with administrators, clients and other interested parties, with the purpose of requesting aid or other types of resources.

Collaboration with *start - ups* is another interesting source of industrial innovation.
The start - up provides flexibility, agility and disruption. In return the companies provide a work space, support services and consultancy, training, mentoring, and a number of tools such as distribution, marketing, sales, corporate governance, etc. so that projects are accelerated. This section includes for example the **BFA** (Business Factory Auto), an initiative in Galicia which aims to promote the incorporation of innovative business to act as a shuttle to highly qualified employment. There are two phases contemplated in this initiative, depending on the degree of maturation of the projects: one of acceleration (for the incipient) and another of consolidation (for more advanced projects). In the first phase, there is a subsidy of

125,000 €, while in the consolidation phase this amount rises to 250,000 €. In addition, the projects benefit from other resources such as: physical space, mentoring, access to leader companies, training, tutoring, etc.

Open innovation and co - creation are other solutions that are gaining prominence as drivers of technological change.
Open innovation creates an environment where businesses and the wider community can actively participate in the creation of mutually beneficial solutions. By interacting with wider stakeholder groups it is easier to solve complex problems and improve processes.

While open innovation suggests an active collaboration between different stakeholders and exchange of intellectual property, **co - creation** develops most commonly between a company and end users, so that they can exchange knowledge and resources.

"Technology is nothing. The important thing is to have faith in people, that they are basically good and intelligent, and if you give them tools, they will do wonderful things with them. "

Steve Jobs

3.3   ROADMAP

It is advisable, before facing major investments in 4.0, to manage the preparation of a roadmap visualizing the progressive changes that we are going to give in each of the 8 areas of value creation of Industry 4.0, among which we should highlight the **people,** especially the *Digital Training*, because it is the basic lever of the digital transformation wheel.

In the first place, it will be necessary for employees and collaborators to know the technology that makes Industry 4.0 possible, the changes it will generate and the benefits of its application. As a natural tendency of the human being, any change or innovation brings with it a rejection; that is why the companies that launch themselves into Industry 4.0 should seek a process oriented towards the construction of **a better environment in the workplace**.

Obviously we could also give the full weight of the transformation of our factory to an external consultancy firm, but there is a risk that we will end up implementing "their solution" and not "our solution". The biggest factor of success is to find the balance between benefit from the experience and expertise of advisers or consultants, and **the involvement of our own team** trained and motivated to create *our factory of the future.*

Secondly, the company must identify the new competences necessary for the implementation of digitization and follow some guidelines for its implementation, with special emphasis on the activation of the talent of our current teams.

At the same time, a **change in organizational model** with a *Digital Enablers Network* inside and outside the company (people with high digital training) to facilitate the transition to the new industrial concept will be necessary.

In this new scenario we start talking more usually about the *Chief Technology Officer* (CTO) or the *Chief Information Officer* (CIO) as an important figure in the organizational structure of companies. As Peter Drucker, one of the fathers of the modern enterprise, says "the best structure will not guarantee results or performance. But the wrong structure is a guarantee of failure. "

Once the digital training has been developed, we will continue with the necessary roadmap to reach the defined objectives. Referencing again the *test of maturity,* we will establish the necessary actions to move forward in each of the areas of value creation. We should not approach this phase in a disruptive and risky way; it is desirable to identify a low budget project or prototype test mode. This project will provide **learning experiences** for us and will show us the possibilities and limits of the enabling technologies of Industry 4.0.

Finally we will reach the phase of **validating** the results and **replicating** solutions to other machines or lines of business.

The route that takes us to Industry 4.0 is really fascinating. Man reached the moon with a much less powerful computer than any Smartphone used today. If we were able to achieve that success with such basic technology, the limits of this new industrial revolution will be only those of our own imagination.

"Never walk on the path traced, it will only lead you to where the others went."
Alexander Graham Bell

3.4   SKILLS NEEDED FOR PEOPLE

Various studies ensure that within fifteen years, out of every hundred jobs the eighty more mechanical will be replaced by machines. In this scenario it is easy to understand that the jobs of the future will require people with skills different from the current ones. What is sought in work now and in the future will be what are known as "**soft skills**"; these are neither knowledge or technical skills, but personal skills. They are therefore transversal skills, which are directly related to the attitude of the worker, and for the **World Economic Forum** there will be:

1.**Solving complex problems** .- The solution begins by making clear the facts, the **data**. And the new industrial revolution stands out precisely because of the large amount of data and information that we will have to process and analyze. In the jobs of the future it is essential to have this skill, to have a good handling of problems, not to block yourself, to see the process not as an obstacle but as an opportunity.

2.**Critical thinking** .- The intelligent critique is indispensable. Having critical thinking does not mean to contradict everyone or not agree with anyone. Critical thinking demands clarity, precision and evidence, and in that way we avoid personal impressions. The process consists of analyzing, putting in order, explaining, schematizing, justifying and checking for application.

3.**Creativity** .- Creativity is a human capacity that machines may never fully develop, and therefore one of the fields in which there will be more job opportunities in the future.

4.**People management** .- This ability is something that requires an understanding of your own personality in addition to understanding the personalities of those around you. It will be necessary to learn to delegate tasks, motivate those around you, always seek to have clear communication with people on your team and outside of it. A company is nothing without its employees.

5.**Coordination with others** .- For coordination is necessary to be able to do two things: make executive decisions and at the same time be open and flexible enough to listen to the opinions of others and to take all other ideas into consideration when taking a final decision.

6..- **Emotional intelligence**.- With emotional intelligence, a person has a higher level of empathy, self - regulation and self - awareness that can work more easily with others. People who have developed emotional intelligence know how to react to others based on their own personality.

7.**Make judgments** .- In future work we will have to choose between multiple options and choose the best and most effective. This will require using reasoning with intuition to analyze each element of a problem and choose the best decision. This skill requires a good state of mental health, a predisposition to face problems (with a patient, tenacious and positive attitude), and self-confidence.

8.**Service orientation** .- This skill is identifying and anticipating customer needs and finding ways to provide service and attention that not only cover those needs effectively but also exceed expectations.

9..- **Negotiation skills**.- An optimal negotiation is based on redefining the problem or conflict considering all affected parties and interests to establish options that benefit all. This skill is based on the development of a sincere, calm atmosphere focused on solutions rather than conflict; a clear and objective , but empathetic way of communicating (you try to put yourself in the place of all parties when communicating); an analysis of the differences that can cause future conflicts; and detecting interests that are not the main objective but that may be opportunities for agreements in the negotiation.

10. **Cognitive flexibility** .- This skill refers to the ability to adapt and approach different unexpected situations. This involves two skills: a person must have the ability to be flexible and adapt to problems that may arise, and that person should also be able to learn from processes and techniques quickly when they enter a new or unfamiliar environment.

Top 10 skills

in 2020		in 2015	
1.	Complex Problem Solving	1.	Complex Problem Solving
2.	Critical Thinking	2.	Coordinating with Others
3.	Creativity	3.	People Management
4.	People Management	4.	Critical Thinking
5.	Coordinating with Others	5.	Negotiation
6.	Emotional Intelligence	6.	Quality Control
7.	Judgment and Decision Making	7.	Service Orientation
8.	Service Orientation	8.	Judgment and Decision Making
9.	Negotiation	9.	Active Listening
10.	Cognitive Flexibility	10.	Creativity

Source: "Future of Jobs Report", World Economic Forum

4.KETS (KEY ENABLING TECHNOLOGIES).

Converting conventional factories in **Smart Factories** requires the development of digital enablers (Technology 4.0 or Key Enabling Technologies).

Next we will look at the main enablers 4.0 from a conceptual point of view and also with practical examples of their current applicability. Industry 4.0 has the ability to transform **data into information** and **information into knowledge** so you can optimize the process of **decision-making** in business. Therefore let's start talking about DATA:

1. Big Data and Advanced Analytics

Big Data is a concept that refers to data sets so large that traditional computer applications of data processing are not enough to deal with them. Data analysis is not something new. For years, companies have tried to take advantage of the information they receive. The problem arises when we face large amounts of data, we are speaking of hundreds of GB, terabytes generated in small spaces of time, because that is when the hard drives of common systems are saturated.

The data in factories will grow vertiginously in the short term due in part to the massive collection of information from wireless sensors and devices such as artificial vision cameras, radio frequency identification readers, etc. In the digital domain everything leaves a 'footprint' and this generates a multitude of data, which will serve to predict behavior, make decisions in advance and achieve maximum efficiency.

Objective to be achieved with BIG DATA -> decisions and actions based more on data analysis and less on experience and intuition.

" It's the decade of data and that's where the revolution will come
from"
Alex Pentland (MIT)

4 V OF BIG DATA, are the four elements that characterize this technology:

VOLUME:

Companies handle hundreds of Terabytes of information (1 terabyte = 1,000 Gigabytes)

VARIETY:

Different forms and sources of data, such as *wearables* (sets of appliances and electronic devices that are incorporated some part of our body interacting continuously with the user and with other devices in order to perform some specific function. Footwear with GPS and wristbands that control our health are examples among many others of this technology that is going to become popular in the short term.

VELOCITY:

Real-time analysis (a modern vehicle has more than 100 sensors, which monitor everything that happens in the environment such as tire pressure, fuel level, presence of obstacles, etc.)

VERACITY:

The information will facilitate reliable decision-making.

Source Mckinsey Global Institute

ADVANCED ANALYTICS PROCESS:

1.**Capture.** The meters and sensors (temperature, light, height, pressure, sound ... that transform the physical or chemical quantities and convert them into data.) They have existed for decades, but the arrival of wireless communications (WI-FI, Bluetooth, RFID, ...) has revolutionized the world of sensors.

2.**Transformation.** Once the sources of the necessary data have been found, the next objective consists of making the data collected in the same place, formatting them and making transformations (data conversions and dirty data cleaning). This is what is called ETL platforms (extract, transform and load).

3.**Storage.** More flexible and allows you to manipulate large amounts of information much faster than traditional databases. SQL (structured relational model) and NoSQL (dynamic and flexible scheme).

4.**Analysis.** For the analysis process three techniques are mainly used:

> **Association:** to find relationships between different variables.

> **Data Mining:** aims to find predictive behavior. It encompasses the set of techniques that combines statistical methods and machine learning.

> **Clustering:** which divides large groups of individuals into smaller groups of which we did not know their similarity before the analysis in order to find similarities between them.

5.**Display.** A picture is worth a thousand words, or a thousand pieces of information. It is easier to interpret data displayed in graphs or maps that other tables with numbers or text.

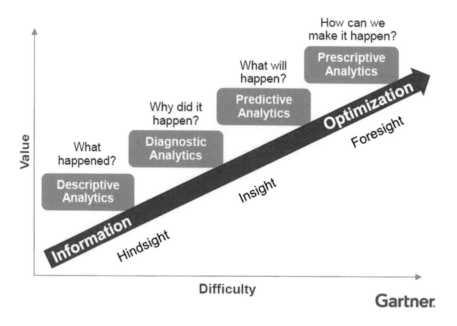

Analytical types. Source: Gartner Business Intelligence Summit

Descriptive analytics (the most basic component of analytics) shows us what is happening in our processes in a way that is easy to understand. Analyzing historical information allows us to determine what has happened and why.

The techniques of predictive and prescriptive analytics are the next step that will help us convert the data into knowledge and decisions. On the one hand, **predictive analysis** allows us to predict what will happen and **prescriptive analysis** helps us identify the most appropriate decisions for our processes.

"Information is the fuel of the XXI century and data analytics the combustion engine"
Peter Sondergaard (Gartner)

Case study 1 : **Google** self-driving car *(Waymo)*.

Google has been working on self-driving technology since 2009. The cars they use for the project are equipped with cameras, GPS, internet connection, and a range of computers and sensors that allow the vehicle to safely circulate on public roads without human intervention.

The *Waymo* is designed to detect pedestrians, cyclists, vehicles, road work and more from a distance of up to two football fields away in all directions. It is one of the biggest challenges of Big Data , as the computer has to process more than 1 million data per second.

Case study 2 : **Amazon.**

The great leader of e - commerce has a Big Data system that allows them to send the product to the distribution center before the customer buys. It is able to predict very reliably the purchasing desires to be produced in their customers, which facilitates a novel delivery system orders in less than two hours after making the purchase *(Prime Now)*. Its application represents a real transformation of the logistic models and is driving the rest of the companies in the sector towards a mandatory transformation that allows them to optimize their logistics management to give the user the immediate response required by the digital world.

Case study 3 : Big Data applied to Transport and Logistics: **Sateliun**

The Logistics and Transport sector generates enormous amounts of data that can be exploited using Big Data technologies. Sateliun combines geolocation tools, vehicle management by GPS and Big Data, to establish more efficient routes, with the consequent saving of time and fuel. It also displays the updated and real-time status of the road network, which improves efficiency, especially in large urban centers. Tracking merchandise is more accurate and more reliable predictions can be made about the demand for a service, so that the company can move forward to the market and manage its workload more efficiently.

Case study 4 : **Mercedes- Benz** (Vitoria-Gasteiz Plant)

Mercedes plant uses optical sensors to collect information on about 200 parameters related to the quality of the paint of each of the 700 vans that leave its assembly line each day. These data are used to predict situations, identify patterns and know how a machine is behaving during the production process. In this case, they help to certify the quality controls of the paint of each of the vehicles.

Sindelfingen factory, the largest production plant of Mercedes-Benz.

Case study 5 : **Michelin Group &** Intelligent tires.

The Michelin tire company markets an intelligent Evolution3 called MEMS incorporating water-resistant sensors that provide information about its temperature and pressure. This tire has 3G connectivity and is capable of sending data in real time and even sends alerts via email or SMS if the temperature or pressure thresholds are exceeded. At the moment, this product is intended for heavy machinery in the mining sector, but the technology could be applied to any type of wheel.

Sources: McKinsey Global Institute ; Gartner ; Baoss [2017]

"It is a major mistake to theorize before one has data. One begins to alter the facts to fit them into the theories, instead of theories to fit the facts. "
Sherlock Holmes.

2. Next-Generation Sensors .

One of the true drivers of Industry 4.0 has been the cheapening of sensor technology, which makes it possible to measure infinite parameters in processes (not only in machines, but also products, storage media, people, environment, etc.).) and therefore allows us to monitor everything that has happened in the factory, what is happening and, as we will see later in the Machine Learning and Digital Twin section, will allow us to even predict what will happen.

No automation without gaps is possible without sensors or actuators. The intelligent connection of sensors and control devices to the control level (PLC) ensures accurate and cost - effective manufacturing.

Sensors are elements that convert physical quantities into electrical signals transferred to the controller. They can be: inductive, capacitive, optical, magnetic, ultrasonic, etc. Their function is to detect presence, level, pressure, temperature, flow, pH, etc... and communicate it to the system.

The actuators execute the command received from the controller and can be pneumatic, electric or hydraulic. They perform a linear or rotary force or movement, or other actions according to specific design. At the level of communication, you can find network topologies with star form or ring form, under open communication protocols (Profibus, AS-I, Control Net) or proprietary code (CC Link), usually under RS-232 communication standards or RS-485.

To implement the concepts of Industry 4.0 in the automation industry, sensors not only have to provide signals or measured values, but they also need to establish fast and reliable communications.

Within the field of communication, IO-Link has emerged with strength, it is a digital interface with which the measurement values are not distorted by interference in the cable. Another great advantage of the transmission with IO-Link is the option to add additional information (such as the status of the sensor) and communicate it simultaneously. In terms of sensor configuration, IO-Link offers many advantages, since parameter settings, for example, can be adopted directly from an IO master. Therefore, a sensor adjustment that requires a lot of time is no longer necessary.

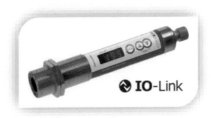

Noncontact temperature sensor, infrared, with IO Link

The most important characteristics are:

Simple and easy wiring

The connection between a master and an IO-link device is allowed up to a maximum of 20 m distance, with an unshielded 3-wire cable.

Easy operation and automatic parameter adjustment

Each IO-link device has an associated IODD (IO Device Description, or input-output device description). This file contains information about the manufacturer, article number, functionality, etc. This information can be easily read and processed by the user.

Unlimited expansion

With a single IO-link port, up to 16 standard bi-stable sensors / actuators can be linked by a three-wire cable of up to 20 meters in length. In this way, elements such as conductors, terminals, interfaces, junction boxes, cable chains, etc., are drastically reduced at the time of installation.

100% compatible

The IO-link consortium covers the world's largest suppliers of industrial automation technology. There are already IO-link devices to connect sensors and actuators to the most popular Profibus fieldbus networks, Profinet, Ethernet I / P, Ether-cat, CC-link, etc.

New applications:

From the point of view of **Logistics**, the cheapening of sensor technology will allow each product or container to carry meters of various parameters and therefore can be monitored.

In the field of **Robotics**, there are new sensors (such as the S-250 from Epson) that will automatically modulate the force that machines apply to objects, allowing the automation of complex tasks that until now required human intervention.

Biomechanics uses sensors distributed by the human body. With them we can measure muscle activity, strength, movements, or even vital signs such as heart rate, etc.

Together with the most classic manufacturers, we have also experienced the birth and subsequent rise of an element of technology that is now essential for many, and that is intimately linked with the Internet of Things: **Arduino**, which is a free hardware company and a technological community that designs and manufactures the most modern sensors.

Until now the electronics was very limited to an almost purely professional field, but the arrival of Arduino has allowed anyone to enter this world. Much has helped the **low cost of the components** and the **huge catalog** of available accessories: from the simplest buttons, ultrasonic, light or distance sensors .; even touch sensors, accelerometers, tilt sensors, potentiometers, humidity and temperature sensors, altitude, pressure, etc.

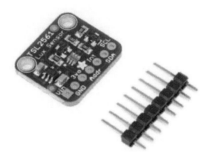

Arduino Light Sensor

A single complex machine can have more than a thousand sensors, which in some cases must be read in very short cycle times (milliseconds), providing millions of readings per year. This shows the magnitude of the level of information provided by the machines and production lines of the Factory of the Future, which implies (as we will see later) not only new ways of handling the enormous amount of information, but also new forms of organization of information so that it can be easily digestible and compressible by human operators and, therefore, facilitate the decision-making task.

Sources: IO-Link; Arduino ; Xataka [2017]

" A lot of times, people don't know what they want until you show it to them."
Steve Jobs

3. IoT

Internet of Things is a concept that is based on the **interconnection of any product with any other around it**. The goal is to make all these devices transfer data to a network without human intervention and, therefore, be more intelligent and independent. This requires the use of the **IPv6** protocol (128 bits). The huge increase of IPv6 in the address space is an important factor in the development of the internet of things: we could assign an IP address to every "thing" on the planet and we still have more than enough.

WHAT ARE THE KEY TECHNOLOGIES OF THE INTERNET OF THINGS?

I. **Small, low-power processors** (a "minicomputer");

II. **Latest generation sensors;**

III. **Low consumption communication.** We already have the data stored in a small computer, but this one is not powerful enough to be able to process them quickly. As a solution, we will move that information to another computer through some communication channel, for example of local network connections via Ethernet or wireless transmission through mobile connectivity or also Bluetooth 4.0 LE (Low Energy) that is thought to be implemented in systems with reduced batteries.

WHAT IS THE DIFFERENCE BETWEEN IOT AND IIOT ?

IIoT devices are **more resistant** since they have to work under extreme conditions without breaking down and are usually located in places of difficult access. For example, they must withstand higher temperatures, be resistant to corrosion and be able to be submersible in water. Therefore, they must contain long-lasting batteries and consume the least amount of energy possible. They must be given **great autonomy**, since sending a maintenance worker to the device is usually very expensive, and have their own control and monitoring systems to prevent early deterioration.

IIoT devices generate a large number of data points (monitored variables) and, therefore, need **capacity** to increase whenever necessary. Sometimes the IIOT device itself performs a preliminary analysis and processing of the data before sending it to the general control system in order not to saturate it. IoT, on the other hand, is not so scalable or doesn't produce so much data.

IIoT devices have their own **security systems**, more robust than those of IoT. A cyberattack could be fatal for a plant or facility, since IIoT is connected to infrastructures and networks crucial for the operation and industrial exploitation.

IIoT devices must have a high capacity for **customization**, since their configuration will depend on the place where they will be placed, the function they will have and what they will be integrated into. In contrast, IoT devices are more "plug and play", are not designed for this degree of customization.

HOW WILL 5G INFLUENCE THE INTERNET OF THINGS?

It is expected that the fifth generation of 5G (wireless broadband) connections will start operating before 2020. These connections are essential to the advancement of Industry 4.0 for the *transmission speeds, high capacities , low latencies and consumption reduction* that it promises.

Speed.

They will be 100 times faster, with average speeds of 20 Gbps per second.

Capacity.

The 5G allows us to improve the efficiency of the use of the frequency band and multiply by 100 the number of connected devices. Currently there are 7000 million devices connected to the Internet. When the IoT becomes widespread, it is expected that there will be up to 100000 million devices connected in 2025.

Latency

It is the response time that a device takes to execute an order from when the signal is sent. The lower the rate, the faster the reaction of the device that we operate at a distance. 5G reduces that delay to 1 millisecond, from more than 10 milliseconds of 4G.

Energy consumption.

The 5G reduces the power consumption of the network by 90%, and allows the batteries of the machines such as alarms or sensors to last up to 10 years.   *Source: Huawei*

Case study 6 : Application of the Internet of Things in the cities through road signs. Applying the IoT, if we go at a speed greater than that which is allowed, our car would automatically reduce speed when receiving the data from any of the signals that surround us. This also would facilitate the arrival and expansion of autonomous transport in our cities.

Case study 7 : PPE (Personal Protective Equipment) with built-in tag could be designed (as we will see later in the RFID chapter), in order to detect if an operator is using the correct PPE with a machine or facility.

Case study 8 : Neteris application for a major industrial equipment manufacturing company.

The objective was to demonstrate that during transport the humidity and temperature parameters necessary for the correct handling of equipment were respected. The first challenge was to find a device that met the following characteristics:

- o Device with control sensors, both humidity and temperature.
- o Small size because it had to go inside an industrial team.
- o No online connection was necessary (maritime transport), but it did allow downloading the stored information once

the coverage was recovered and, moreover, internationally.

- ○ Reusable.

2. Information storage .

The next challenge was to define where to store all the raw information generated by each device. In this case, it did not apply to maintain that information in the ERP, so it was decided to store it in a cloud platform with full availability to the user.

3. Information integration.

The key information generated by the device and related to each team was integrated via BSSV (Business Service) from the cloud platform with the Oracle ERP solution, allowing the user to know at all times what equipment could have incidents in its transport or handling.

4. Traceability

Assigning the serial number of the device with the serial number of the manufactured equipment by bar code reading

Humidity and temperature sensor -> Communication -> Cloud Platform

This solution allowed the customer to monitor and control the quality of transport and handling of the equipment by the transport companies. In addition, it was able to demonstrate with reliable data when the chain breaks occurred.

Sources: Inmarsat ; Techt arget ; Oasy s ; Neteris . [2017]

"The only way to get good ideas is to have lots of ideas"
Linus Pauling, Nobel Chemistry and Nobel Peace Prize

4. **RFID Tags** (Radio Frequency Identification) and **RTLS Systems** (Real Time Location System)

RFID tags are small devices, similar to a sticker, that can be attached to or incorporated in any object.

They contain antennas to allow them to receive and respond to requests by radiofrequency from an RFID transceiver. Passive tags do not need internal power, while active tags (which send the signal to a reader every few seconds) do require it. One of the advantages of the use of radiofrequency is that direct vision between transmitter and receiver is not required. *Bar codes*, unlike RFID, have disadvantages such as the small amount of data they can store and the impossibility of being reprogrammed.

This technology, together with the **EPC** (Electronic Product Code), will make possible the tracking of products allowing total visibility along the supply chain. The *Electronic Product Code* is a unique number that is recorded on the chip contained in an RFID tag and placed on each product, which allows accurate tracking of each physical unit.

Real-Time Location Systems (RTLS) are used to automatically identify and track the location of objects or people in real time, usually within a building or bounded area. The fixed reference points receive wireless signals from the RTLS Tags to determine their location.

In addition to RFID, there are other technologies on which RTLS can be based:

- **Infrared** (IR). They require a clear line of sight for labels and sensors to communicate, so if a plate is covered or flipped over, the system may not work properly.

- **Ultrasound**. Ultrasound, as a communications protocol, is slower (with longer wavelengths) than infrared, so it generally cannot match the performance of other technologies.

- **Wi - Fi** . Although the Wi-Fi infrastructure is often pre-existing in the performance environment, the accuracy is limited to up to 9 meters, which makes its value as a location tool uncertain.

- **UWB** : The advantage of UWB technology is the high level of transmission security. The UWB signal is difficult to detect and locate, because the spectral power density is below the thermal background noise. It can reach a precision of 10 centimeters at measuring distances of up to 100m.

- **BLE** : *Bluetooth Low Energy* appears from the specification in version 4.0. It is aimed at very low power applications powered by a button cell battery. It has a transmission and data transfer speed of 32Mb/s. It works on the 2.4 GHz frequencies and was created for marketing reasons for smartphone and tablet devices. Important advantages of this technology are that it is based on a universal standard, and is immediately available on mobile devices without the need for hardware.

Source: vmvicente [2017]

Case study 9 : Smart Key of a car.

The key employs an active RFID circuit that allows the car to recognize the presence of the key within one meter of the sensor. The driver can open the doors and start the car while the key is still in the wallet or pocket.

The same could be applied if we want a machine to be activated in the presence of a qualified operator and prevent untrained personnel from starting up an installation.

Case study 10 : Safety of the personnel in the Factories.

Monitoring, for example, the presence of maintenance personnel by means of bracelets and preventing an installation from being started accidentally, with the presence of some operator inside the fence of a robotic cell.

Another possible application is the monitoring of people inside hospitals.

Case study 11 : Smart **Warehouses**.

Currently, the most important application of RFID is logistics. Using this technology offers the possibility to have any product within a store or even within the supply chain located.

KALEIDO , IDEAS & LOGISTICS has designed an automated geolocation system to help the processes of loading, unloading and / or storage of irregular topology merchandise.

With CARGO LOC you can anticipate the best arrangement of the goods in the warehouse. This information is transmitted to the mobile devices of the operators (PDAs), so that they can monitor the operations and validate the disposition of the goods in the previously assigned site. The final site is registered by GPS and can be consulted from the office through a specific application. Thanks to RFID and bar codes, CARGO LOC is also able to manage warehouses in open spaces.

Case study 12 : **Bosch Track and Trace** (with the collaboration of Cisco, SAP and TechMahindra) .

Fleet management of power tools :

In the construction of aircraft, there are precise rules that specify the type of screw and the amount of force that must be used to join the different components. When it comes to passenger planes, there are thousands of screws that must be tightened and documented accurately. The joints in the wings, for example, require a different amount of force than those in an airplane window.

The **Track and Trace** tools will be able to identify your precise location in the workshop. With the position of the aircraft also fixed, we can determine that a particular tool is in production and then we can send instructions that specify the force you must use to tighten the screws there. In addition, if a tool recognizes that it is being misused, it will shut down to prevent accidents or injuries. And it will also collect data on its use and status. *Source : Bosch [2017]*

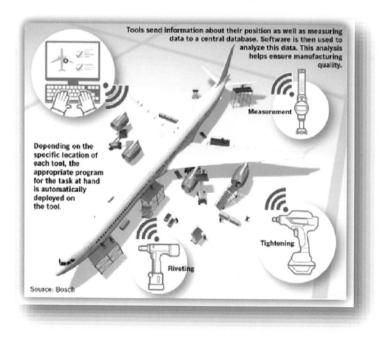

38

Case study 13 : Situm RTLS

Indoor positioning technology:

Situm RTLS is the name of a new location technology for smartphones. It works like a GPS inside, which allows locating people and objects anywhere.

Situm RTLS calculates a new position estimate every second. However, the working frequency can be set up to a maximum of 10 times per second. Its margin of error is very small (1 m) due to the fusion of information from different sensors: Wi-Fi, magnetometer, Bluetooth Low Energy (BLE) and prediction of user movement based on the inertial sensors (accelerometer, gyroscope). It is also able to automatically detect at what level of the building we are.

Indoor location and navigation systems are useful in many areas, such as:

• Indoor location and turn-by-turn navigation.

• The investigation of the user or client. The analysis of shopping habits: heat maps, workshop time, market studies, movement patterns, etc.

• Resource management. Control of routes and information about the situation of employees and security team in real time.

Another possibility that is offered to us is step-by-step navigation, which consists of guiding a person to a specific point including instructions to follow a route from their current position. This guide can also come in the form of audio instructions.
It is also possible to collect information from visitors to a space. With these data you can study the habits of customers or employees.

Sources: Situm ; Bosch; Industrial Internet Consortium ; Kaleido [2017]

5. Automatic Guided Vehicles (AGV).

AGV defines those vehicles that move without driver.

This type of vehicle is being implemented in industrial environments such as logistics warehouses or assembly plants to perform very repetitive routes transporting goods from one point to another (or several intermediate), or to work in environments that are complicated for human beings (due to high toxicity, extreme temperatures, etc.). The idea is to simplify this task of distribution and collection with a fully automated vehicle, which follows a schedule established in advance, or performs a non-predefined route guided by signals (inputs) that indicate on the fly.

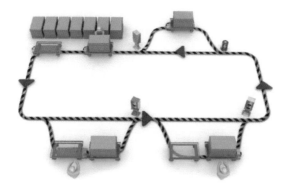

AGV logistics management system

Depending on the guidance systems and their automated control, we can talk about autonomous systems, which are not related to the elements of their environment; or complex systems, those that receive inputs from the elements of their environment and act conditioned by those signals.

The different guidance systems used today are basically the following:

- Guided by thread. A thread buried in the ground is used to mark the route to follow.

- **Magnetic**, that uses magnetic bands or tags in the tracing of routes, is the most advisable to maintain a flexible layout).

- **Artificial vision.** Tracing routes by painted band thanks to the installation of a camera on board the AGV.

- **Guided by laser.** Through a unit that maps a mesh of reflectors established in the environment.

AGV designed by SSI Schäfer

Attending to a second classification, we can speak of vehicles of standard manufacture that are adapted to work in an automated way - although they can work alternatively in a manual way -, by means of a system of navigation and management to the measurement; or specific manufacturing equipment to work as AGV. Each of them has its advantages and disadvantages, so depending on the needs, one or the other will be determined.

ASTI AGV manufacturer, Spanish production company leading in Europe

For a purely logistical use such as a warehouse where there is little human traffic, it is possible to develop a system of vehicles that move more quickly, which would favor the choice of standard manufactured forklifts converted into AGV. If it is a matter of delivering material in an automotive assembly line, or distributing small loads within a hospital, where there are greater limitations due to constant human interaction, it is preferable to install a *guided by thread AGV* with fixed routes and lower speeds.

For the **battery recharging** system there are the options of: *manual exchange; automatic; of opportunity* (the latter allows the partial recharge of batteries in the moments of inactivity of the system).

An operator performs guided and charging test with an AGV

The benefits for a company are evident, in terms of increased security, greater productivity, elimination of human errors, stability in delivery routes, and reduction of costs in the medium term. They are systems with a certain economic cost, but for companies that work three shifts, the return on investment occurs very quickly.

Sources: ASTI; T. Calatrava; SSI Schäfer [2017]

AGV SSI Schäfer delivers goods in a post-production.

" Where there is a successful business, someone has sometime made a brave decision. "

Peter Drucker .

6 HMI: Human-Machine Interfaces.

HMI is the device or system that allows the interface between man and machine. In recent years, touch technology is the dominant one (the physical keyboard is retreating). The most common types are, on the one hand, *capacitive pads* not sensitive to pressure. On the other hand, *resistive tactile pads* very useful to work, for example, in warehouses, where the use of thick work gloves is required.

New HMI technologies such as multi-tacit interaction or the generation of contextual menus, categorized by user roles or production states, are new approaches that are beginning to be integrated into some of the catalogs of machine tool suppliers. At present, advances are being made in HMI technologies without contact (Touchless).

Case study 14 : **DanobatGroup** HMI .

To improve the man-machine inter-relation, this Basque manufacturer has developed an HMI device that simplifies the operation of the machine, incorporates programming aids and specialized machining cycles, facilitates maintenance tasks by reducing unproductive times, and provides information to minimize the energy consumption.

Sources: Indusoft ; Siemens ; Danobatgroup [2017]

7. SCADA (Supervisory Control and Data Acquisition).

SCADA, acronym for *Supervisory Control And Data Acquisition*, is a concept that is used to make a software for computers that allows us to control and supervise industrial processes at a distance. It facilitates feedback in real time with field devices (sensors and actuators), and controls the process automatically. It provides all the information that is generated in the production process (supervision, quality control, production control, data storage, etc.) and allows its management and intervention.

SCADA systems traditionally have direct signals for communications, but it is also possible to communicate via wireless (for example if we want to send the signal to a PDA or a mobile phone) and thus not having to use cables.

Often the words SCADA and HMI induce some confusion. The difference lies in the **monitoring function** that SCADA systems can perform, to observe from a monitor the evolution of the variables that we are controlling.

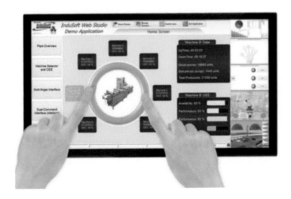

SCADA InduSoft Web Studio®

Case study 15 : Andon panels

that show the KPIs and anomalies in a position or in a line. Acquiring the information with a SCADA we can monitor the frequency of incidents that occur and assess the level of preventive maintenance we have.

Case study 16 : SCADA for an **Oven.**

Control of the operation of an oven manufacturing line.

We manage the entry of raw material by weighing it into an electronic scale connected to the PLC. Then, the furnace temperature is regulated by the internal PID of the PLC modulating the opening of the valve of a gas burner in order to keep the set point entered. The rest of the elements of a line are also controlled: transport, drying, etc. automatically.

The whole system is managed from the SCADA connected via Ethernet. It allows us to change all the set points of temperature, time, weighing, etc. We can visualize oven temperature graphs, generate daily data tables of the most significant values of the process.

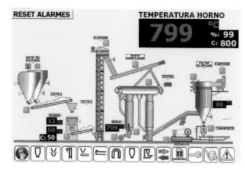

Source : ACESA [2017]

Case study 17 : **Meteorological stations** managed by SCADAs within the factories allow managing control, supervision and activation of heating systems, air extraction, lighting, etc.

8. MES (Manufacturing Execution System).

The MES system is a software tool that functions as an extension of the ERP (Enterprise Resource Planning) system, but oriented to the planning and execution of production. In this way, while the ERP system determines what has to be manufactured, the MES system provides the necessary functions for the management of key areas in plant, such as people, materials, processes, quality, traceability, maintenance and visual factory. Without papers, with precision and in real time, in MES allows the automatic capture of production data, the monitoring of operations and the identification of waste and area of performance improvement (OEE) and lead-time.

Case study 18 : Captor .

The MES system developed by **Sisteplant** is one of the most powerful on the market and is "designed to provide technological intelligence to the decision-making process in the plant".

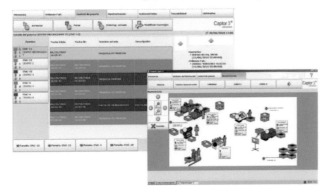

Case study 19 : Q- Plant

The MES system developed by **TAI** (Barcelona) coordinates the integration with the management systems (ERP) and the machine signals (PLC or others) thus having the necessary information for the monitoring and control of the plant activity.

Sources: Sisteplant ; INDA; TAI [2017]

9. **CMMS** *computerized maintenance management system.*

It is a software tool that helps in the management of the maintenance services of a company. The objective: to manage with the CMMS the *corrective, preventive, and predictive maintenance* processes in the production equipment, but also to control *stocks, purchases, and contracts*. With this model it is possible to have a *global vision of maintenance performance*, measuring the efficiency of the maintenance department in each line, with costs, stop times and intervention times.

Case study 20 : Prisma .

The Maintenance Management system developed by **Sisteplant** allows a progressive implementation thanks to its modularity and meets the main global standards such as PAS 55 (ISO 55.000).

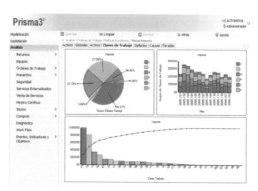

Case study 21 : LINX

The Maintenance Management system developed by **SPI** (Barcelona) CMMS_Linx 7.0 manages the areas of Assets, Human Resources (internal / external), Spare parts (management, stocks, orders.), work orders (Preventive, Corrective, Predictive.) , Intervention Requests, Statistics, etc.

Sources: SPI; Sisteplant [2017]

10. Additive manufacturing.

3D printing is an additive process that allows you to create objects layer by layer from bottom to top. The precise technology depends on the materials, the aesthetics, the mechanical properties and the performance you need. There is the possibility of manufacturing from a low cost prototype to a final model of high resolution and quality with properties similar to the final product. The most striking feature of this technology, unlike machining processes, for example, is that *the more complex the design, the more economical it is to manufacture*.

EXISTING TECHNOLOGIES:

Plastics material deposition technologies:

Fusion Deposition Modeling (**FDM**). It is a technology that allows parts to be obtained using ABS plastic or PLA (a biodegradable polymer that is produced from an organic material).

3D printing technologies with laser:

Stereolithography (**SLA**). Technology based on the solidification of photosensitive resins layer by layer by means of a laser with ultraviolet frequency. The resulting pieces are of high precision and the materials can have different finishes (transparencies, metallized, painted).

Selective Laser Sintering (**SLS**). It uses powder material (polystyrene, ceramic materials, glass, nylon and metallic materials). The laser impacts the powder and melts the material and solidifies (sintered).

Other Technologies:

PolyJet photopolymer. A liquid photopolymer is ejected and then solidified by ultraviolet light. As with all other 3D technologies, printing is done layer by layer.

Selective Laser Melting (**SLM**) that is similar to SLS but that melts the powder material instead of just melting it at low temperature.

Electron Beam Melting (**EBM**) that uses an electron beam instead of a UV laser to melt dust.

Laminated Object Manufacturing (**LOM**) where different layers of material (adhesive paper, plastic or metal sheets) are placed on top of each other and glued with resin / glue and cut with the appropriate shape with laser. This process is partly reminiscent of the mode of manufacture of carbon fiber.

Heat exchanger for the automotive sector manufactured with 3D printing by the German brand EOS.

TYPES AND CHARACTERISTICS OF PRINTERS:

In the industrial world printers are divided into plastic and metal, which although similar, are two totally different universes.

When evaluating a 3D printer we should take into account different features and factors that help us better understand its benefits and its price.

• **Maximum size of printing area**: The average maximum size of printing area that printers of low and medium cost on the market can print would be about 200mm x 200mm x 200mm. The printers of industrial use, much more expensive, can print sizes larger than 900mm x 600mm x 900mm.

• **Resolution**: It is the equivalent to the resolution of a 2D printer, in the XY plane and is defined by pixels per inch (ppi). Today there are printers with resolutions on the order of up to 600 ppi.

• **Precision**: It is the thickness of the thin layers of material that the printer is adding. The typical resolution is 0.1mm and there are already printers with up to 16 microns (0.016mm) per layer.

• **Printing process time**: Printing speed is a key criterion since a 3D printing can take several hours. In addition, it must be taken into account that there are also printers that use support material that needs a post-processing time to remove it from the printed piece. This additional material is used along with the main one in models and parts that require support to be able to print.

• **Materials**: It is very important in a 3D printer to select the type of material you can use to print, you have to evaluate the different ranges (plastics, resins, polymers, polyamide, ...) and the properties of these materials (nuances, transparency, resistance, rigidity and flexibility).

• **Printing expenses**: You must take into account the price of the material with which it is printed and the price of the spare parts such as heads, extrusion nozzles or trays that must be changed in some models.

Case study 22 : Mold cooling

The metal molding tools used for injection molding contain channels for cooling the mold. With traditional tool making methods, the cooling channels are drilled in the tool in straight lines. The additive manufacturing of metals allows us to design and manufacture the cooling channels when building the mold. In this way, the cooling performance is improved, the life of the mold is lengthened and the leftover material is reduced.

Case study 23 : Aerospace Sector.

Additive metal parts are used in the aerospace sector for functional parts, such as turbine blades, fuel injection systems and blades. Optimizing 3D printing improves functionality and reduces weight. Some lighter parts, consequently, will mean lower fuel consumption.

*Robotic arm "Demonstrator" prototype of the company **Stratasys**, designed to manufacture large light parts for Boeing and Ford, in thermoplastics with mechanical properties common in the aerospace and automotive industry.*

Case study 24 : **Mercedes- Benz Trucks .**

Mercedes Benz in Stuttgart (Germany), manufactures spare parts with 3D technology. Initially they were plastics for the vehicle cabin, but now they are more complex pieces (like the thermostat in the image) with demanding quality tests, which have shown that they have even greater resistance to impact and heat compared to an injected piece.

Sources: Lupeon; 3DNatives; Xataca ; CaminsTECH ; Stratasys [2017]

Tip: 3D printing technology advances very fast, and sometimes it may be more advantageous to have a partner for prototyping until a very defined use for the process is discovered and then we can take the next step in a more consolidated way.

" If I have seen further it is by standing on the shoulders of Giants "
Isaac Newton in 1675.

11. Augmented reality.

This consists of a set of devices that add virtual information to the existing physical information, that is, add a virtual synthetic part to the real one. Augmented reality (RA) is different from virtual reality (VR), in which the user isolates himself from the material reality of the physical world to immerse himself in a totally virtual environment. Through augmented reality we incorporate data and digital information in a real environment, through the recognition of patterns that is made through specific software.

Sources: Itainnova ; iAR [2017]

Case study 25 : HUD (head-up display)

In 2015 **Digilens** presented at the CES in Las Vegas, a system capable of being integrated into a BMW motorcycle helmet to provide relevant information to the driver without him taking his eyes off the road to look at the instrument cluster.
Continental has also developed this technology in the car with the aim of showing relevant information for the superimposed driving that the driver sees "live" through the windshield.

Case study 26 : Caterpillar Maintenance machines

Caterpillar has developed a new application with an augmented reality system for the maintenance and repair of machines in situ. The usual process that the technician carries out to diagnose and repair, consists of moving to the place of the intervention, reviewing the procedure on paper or on a PC and finally executing the repair.

This solution is based on a program that explains the process of how to make a diagnostic inspection and repair the machine step by step, using a tablet, smartphone or augmented reality glasses. The glasses are the best option, since they free the hands, allow you to manage the program with your voice and avoid having to touch the Smartphone with greasy hands.

Another function of the program is to take pictures during the intervention to verify if it is being done correctly. For example, in the replacement of a hose, the operator was not assembling it bent in the correct way. The program was able to detect this from a photo, allowing the operator to correct this point immediately. This function extends the possibilities of use towards the training of new operators.

The more complex the machine and the more laborious the repair process, the greater the advantages of quality and the reduction of intervention times with this type of systems. This technology can also help to avoid many trips of maintenance technicians from the manufacturer, making the diagnosis and even some repairs in a collaborative way between the customer and the manufacturer. This becomes a value for the user, as it can reduce the time of the machine out of service.

Source: Caterpillar [2015]

Case study 27 : VW **MARTA** Application

The German car brand Volkswagen has developed an application installed on a tablet that allows you to recognize the focused electrical cabinet and displays the virtual labels that identify each of the elements on the screen.

Volkswagen has also developed an application that helps its mechanics to know step by step how to fix their cars. Its code name is MARTA ("Mobile Augmented Reality Technical Assistance").

Thanks to the iPad camera or the glasses, the engine can be seen and augmented reality does the rest, showing us the steps we must follow depending on the repair that the mechanic wants to make.

Example of Application of Augmented Reality in Mercedes Benz, visualizing how a motor fits into a chassis

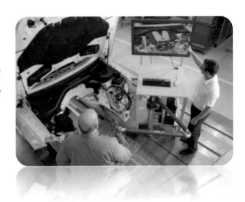

Source: Mercedes Benz [2017]

Case study 28 : Plant Tour with **KPI** display.

By means of Augmented Reality it will be possible for a supervisor, plant manager, director, etc. to walk through the plant and visualize the different statistical data in real time as well as possible faults that an installation may have.

Case study 29 : **DHL** Warehouse.

DHL and **Ubimax**, expert in wearable solutions, are implementing "picking through vision" in the operations of a warehouse in the Netherlands. The graphics displayed on the smart glasses guide the staff through the warehouse to speed up the picking process and reduce errors.

The project has shown that augmented reality generates great added value in the field of logistics and improves the efficiency of the picking process by more than 25%. *Source: DHL [2017]*

Case study 30 : **IntoSite** application from Ford and Siemens for virtual navigation in manufacturing plants.

Ford Motor Company is testing a new **Siemens** software that facilitates virtual navigation within its assembly plants. With this technology the company improves global communication and production.

The new IntoSite application is a **Tecnomatix** web application based on the cloud and uses the **Google Earth** infrastructure. IntoSite has a 3D version of the plants and allows users to navigate virtually through the factories - until they reach the jobs - obtaining a better understanding of the global processes.

In any virtual place, engineers and other team members can add pins - just as they would in Google Maps - and upload content such as videos, documents and images to these pins. This creates a virtual private space where users can easily save and share materials, helping to improve communication within plants around the world.

The IntoSite pilot program is initially under study at the Ford plant located in Wayne, Michigan.

Sources: Siemens; Ford; [2017]

12 Efficient Energy Technologies.

Optimal energy management is key to increasing the competitiveness of companies in the face of the growing increase in the cost of supplies. In industry, the efficiency in the final use of energy is of strategic importance, since *it represents 27% of the world's energy consumption*; this optimization involves the implementation of energy management systems.

The implementation of an **energy management system** allows the measurement history as well as the calibrated energy consumption models of the equipment to be obtained, from which we can establish efficient reference consumption patterns.

With **real-time measurements**, the energy management system is able to compare the energy consumption by process or production line with the reference energy model, which allows us to identify consumption patterns and identify the possible degree of inefficiency that is being operated with. If inefficiencies are identified, the system proposes corrective action measures.

The efficient management of energy in the future will not only be used to control the demand, but also for **generation of energy** in the industry itself through renewable sources (photovoltaic and thermal solar panels, mini-wind) that balance the curve of energy demand and reduce the dependence on the network, as well as smoothing the peaks of consumption with energy accumulation systems (for example, lithium-ion batteries), which allow the use of renewable energy beyond the hours of generation.

The objective is to be **more efficient by recovering** residual heat; the optimization of cogeneration and trigeneration systems; the optimization of the use of raw materials, the reduction of waste and the reduction of environmental impact, etc.

Sources: EnergyLab ; CO2 Smart Tech ; Conexiona [2017]

Case study 31 : **Co2st-tem** System

The "co2st-tem" system of the company **CO2 Smart Tech** is a software for energy monitoring, analysis and management applied to industrial processes and services. It provides a comprehensive approach to energy management, which facilitates the identification, control and optimization of factors related to energy savings and resources (electricity, water, gas, steam, ...). The software addresses energy management from a comprehensive perspective.

Any supply, quantity (pressure, level, flow, temperature, ..) or data (production, units, users, ...) of an installation can be treated and analyzed to optimize its exploitation and its energy efficiency.

Case study 32 : **iPlace** System from CONEXIONA

iPlace is a solution developed for industrial environments. This technology allows remote monitoring and control of electrical consumption and alarms generated in a machine or installation.

The operation consists of the installation of a small plc in electrical panels or in the equipment itself, which is communicated via Ethernet or Wi-Fi with free software that allows us to remotely manage all the systems that make up the facilities:

- Start machines before the usual working day (for example thermal processes that require a previous operating time until reaching the optimum working temperature);

- Detect residual energy consumption in stopped installations;

- Shut down machines remotely;

- Receive alarms caused by an unforeseen machine stop (for example, in installations that operate 24 hours a day, 7 days a week)

13. Collaborative Robots (COBOTS)

The latest advances in robotics technology and the miniaturization of electronic components and processors has allowed the birth of a new era in industrial automation: Collaborative Robots.

They are characterized by being lightweight, flexible and easy to install, and are designed to interact with human beings in a shared workspace without the need to install security fences. The current regulations regarding industrial robotics systems are based on the **ISO 10218-1**, **ISO 10218-2** standards. The emergence of these new Robots has led to new legislation such as **ISO / TS 15066** where security requirements for robots and collaborative applications are defined.

The Collaborative Robots incorporate advanced force control systems for consumption in the axes, making it possible for the robot to stop when it encounters an obstacle, which allows collaborative Human-Robot work without the security fence, as long as the application allows it. **Collaborative robots do not guarantee by themselves that the final application is collaborative**. This depends on many more factors related to the process. But if the robot is not collaborative, the application will not be.

Their **small size**, their **versatility** and their **affordable price** differentiate them from traditional industrial robots and make them suitable, not only for the large company but also for SMEs, who need cost-effective and easy-to-use solutions if they want to take advantage of *Industry 4.0 & Smart Factories* opportunities, and that's where the Cobots make the difference.

One of the benefits is that they are very **easy to program**. In fact, some are even self-taught and can learn a new movement simply with a technician to guide them. Another benefit is that they can be moved and adapted to multiple locations in the production line.

Robotic line in an automotive plant

Human / Machine Collaboration:

In a study carried out by the **BMW** brand, it was shown that teams composed of people and robots that collaborate efficiently can be more productive than teams composed of people or robots alone. **Mercedes Benz** is moving to what it calls "Robot Farming": equip workers with a series of small and lighter collaborative robots.

The strengths of humans are the weaknesses of robots and vice versa. The most interesting thing about collaborative robotics is the combination of skills, taking the best of each one:

Human: *Problem solving, Dexterity and Flexibility.*

COBOT : *Precision, Strength and Resistance.*

When designing a production line with joint work Operator / Robot, it is usual to follow the premise that repetitive tasks are assigned to the robot and the most complex tasks are assigned to the operator. The movements of the robot will be limited in a limited area to avoid interfering with the operator. One of the characteristics of the cobots in that its scope (range of action) is less than that of traditional robots.

Sources: ABB ; Indrobots ; Sisteplant . [2017]

In 2015, more than 240,000 robots were installed worldwide.

The International Robotics Federation anticipates that there will be around 1.4 million new generation industrial robots in service in factories around the world by 2019.

Most cobots are single-arm robots, although there are also some double-arm models.

Main collaborative robots (Cobots) of 1 arm:

FANUC

CR-35iA is the collaborative FANUC robot compatible with the iRVision vision system

Model : CR-35iA

Country : Japan

Web : www.fanuc.eu

Makino Machining center with CR35iA

KUKA

It has 7 axes and a maximum range of 800mm with IP54 protection

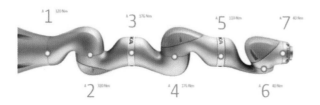

Model : LBR iiwa LIGHTWEIGHT ROBOT

Country : Germany

Web : www.kuka-lbr-iiwa.com

RETHINK ROBOTICS

It has 7 axes and a large reach with little arm weight.

Model : Sawyer

Country : US

Web : *www.rethinkrobotics.com*

UNIVERSAL ROBOTS

The philosophy of Universal Robots is to make robotics accessible to everyone, and in particular to facilitate the implementation of robotics in SMEs. It really is one of the most recommended options.

The approximate cost of the UR ranges between 16,000 and 25,000 euros, which added to the cost of the application would suppose a total of between 50 and 60 K €, which makes them an option with a fast ROI (Return on Investment), when compared to solutions with traditional robots whose total costs approach 100 K €.

Model : UR3 - UR5 - UR10

Country : Denmark

Web : www.universal-robots.es

Remarks:

- **Universal Robots** is recognized in a general way as the pioneer in introducing collaborative robots in the industry.

- **Kuka** and **Rethink Robotics** are the only ones that offer 7-axis cops, which give them greater flexibility than 6-axis coaches.

- Most cobots have a limited maximum load, generally not exceeding 10-15 kg. The ranges are also moderate, between 500 and 1300 mm. There is a notable exception: the **Fanuc** CR 35iA reaches up to 35 kg, with a maximum range of more than 1800 mm (it is in fact a modified M20iA / 35M), although the resulting weight of this cobot (almost a ton) in practice limits its mobility or frequent changes of location.

- **ABB** is not included in the list of manufacturers of 1-arm cobots. Despite having acquired the firm **Gomtec** in 2015, manufacturer of the Roberta model, ABB is not actively promoting it and is focusing its collaborative offer on the YuMi double-arm model (we will see it later).

Other examples of 1 arm cobots:

FRANKA - Emika

Franka Emika is a collaborative robot of the company CKBee of 6 axes.

Model : FRANKA - Emika

Capacity : 3 Kg

Country : Germany

Web : www.franka.de

GOMTEC

Model : Roberta

Capacity : 4, 8 and 12 kg

Country : Germany

Web : www.gomtec.de

Roberta has 6 axes and has been developed specifically for all companies that focus on flexible automation, in particular, small and medium enterprises. An agile and lightweight robot that could easily be moved around the production plant.

KINOVA

Portable and light. It has 6 Axes and only weighs 5.7 Kg and a range of 90 cm. It also has a 3-finger clamp with independent movements and a lot of precision. It can work with water since it has an IPX2 protection.

Model : JACO

Capacity : 1.5 Kg per arm

Country : Glen

Web : www.kinovarobotics.com

MABI

Model : SPEEDY 10

Capacity : 10 kg

Country : Switzerland

Web : www.mabi-robotic.com

PRECISE AUTOMATION

The PF400 is the first collaborative SCARA robot. It has 4 axes.

Model : PF400

Capacity : 1KG

Country : US

Web : www.preciseautomation.com

"Creativity requires having the courage to let go of certainties "

Erich Fromm

Main collaborative robots (Cobots) with 2 arms:

ABB

Model : YuMi

Country : Switzerland

Web :
new.abb.com/products/robotics
/ Yumi

RETHINK ROBOTICS

Model : Baxter

Country : US

Web : www.rethinkrobotics.com

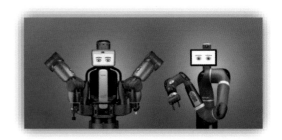

Baxter can handle a wide range of repetitive production tasks, including packaging, lifting material, loading, machine tools. The line workers themselves can train Baxter in a matter of minutes, with no software, robotics or engineering experience required.

Baxter has an LCD screen, sound 360º and 3 vision cameras, force detection, and two arms with 7 degrees of freedom each. It also offers a series of accessories such as vacuum cups and parallel grippers.

Other examples of 2 arms cobots:

PI4

Model : Workerbot3

Capacity : 5- 10Kg

Country : Germany
Web : www.pi4.de

ROLLOMATIC

Model : NEXTAGE

Capacity : 1.5 Kg per hand

Country : Switzerland

Web : www.rollomatic.ch

NEXTAGE is the collaborative robot Rollomatic, 2 robot arms with a human - like geometry designed to perform tedious work. The robot has four video cameras (two in the head and one on each arm) .

It has an accuracy of 0.5 mm and a repeat of 0.03 mm.

YASKAWA MOTOMAN

Model : Dexter Bot

Capacity : 5kg

Country : Japan
Web : www.motoman.es

Sources: AER-ATP; UR; ABB; Kuka ; Infoplc ; CKBee ; Motoman ; Rethink Robotics; Rollomatic ; Mabi Robotic; pi4; Precise Automation; Indrobots ; DFKI GmbH Robotics Innovation Center . [2017]

Case study 33 : Lear Corporation (Germany).

Lear, a global supplier in the automotive sector, has optimized the just-in-time assembly (JIT) by integrating cobots into its production line. The robotic arm performs bolting operations on the seats and digitally supervises the process to prevent defective seats from continuing on the conveyor belt. The result is an increase in both the speed of production and the reliability of the product. Control and flexibility: The company managed to maintain control over the decisions made by the robot thanks to a very simple programming. All electronic elements and controllers are combined within a central point, which allows them to schedule changes without the help of external experts.

What Lear Corporation needed was a small portable robot that could work side by side with the template without safety guards so that the rest of the production phases would take place one after the other on the conveyor belt.

The robot in question should also be easy to program for people with little experience in robotic technology.

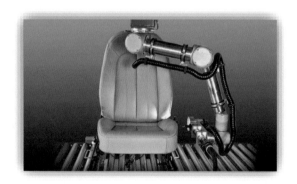

Solution:

Lear opted for UR5 technology due to space limitations.

The robotic arm is responsible for screwing the seats of the car to the rest of the body with a screwdriver at the end of the arm, performing about 8500 operations a day.

The seats are equipped with a transponder that contains the individual identification data. As soon as the robot receives the seat, the transponder is read and the robot tightens the screws on both sides of the seat. If a seat has lost the screws, the robot discards this product and issues a warning signal.

Thanks to the intuitive user interface of the UR5, any employee can program the robotic arm through a sequence of commands on the screen or, simply, grabbing the arm to trace the series of movements that it is desired to memorize.

Source: Lear Corporation [2017]

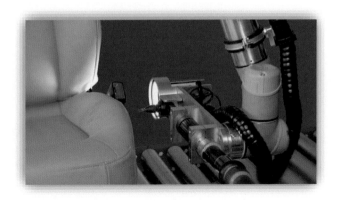

" It is not possible to solve today's problems with yesterday's solutions " . Roger Van Oech

Case study 34 : Continental Automotive Spain (Rubi, Barcelona).

In 2016, the company decided to purchase several UR10 coves to automate the manufacture and handling of PCB boards, and reduced the change times by 50%, from 40 to 20 minutes, when compared to the task performed manually.

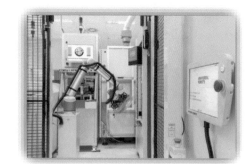

Main advantages:

Control and flexibility: The company managed to maintain control over the decisions made by the robot thanks to a very simple programming. All electronic elements and controllers are combined within a central point, which allows them to schedule changes without the help of external experts.

Less load for the team: The arrival of the cobot was a change in the role of operators, who no longer have to perform non-specialized tasks, such as moving components from one station to the next. Now they can concentrate on specialized tasks that help improve production.

Cost reduction: Automating the work of moving parts and components around the plant has allowed Continental to reduce operating costs significantly.

Greater Security: The Continental team is satisfied with the safety measures associated with collaborating robots. For example, the operator can enter the cell at any time and the robot stops immediately thanks to additional sensors that suspend it when an operator approaches the robot.

Source: Continental [2017]

Case study 35 : **Boco Böddecker** (Germany).

Böco Böddecker specializes in locking and insurance systems for the automotive industry. These plastic and metal parts are used to secure doors and seats in cars from a wide range of manufacturers.

The challenge was to label parts unitarily because their automotive customers demanded that each of the pieces be marked individually with a code. The automation of this simple and repetitive task was a great economic saving.

A cobot was installed that at the same time also labeled quality controls, so the possibilities of delivering a wrong product to a customer have been greatly reduced.

Source: Boco Böddecker [2017]

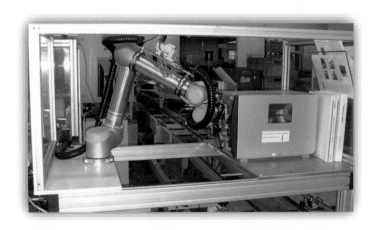

" Innovation is what distinguishes a leader from others"

Steve Jobs

Interesting statistics:

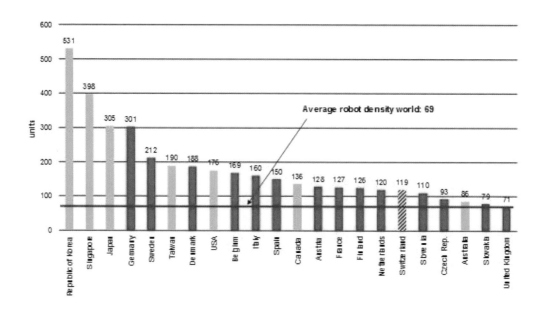

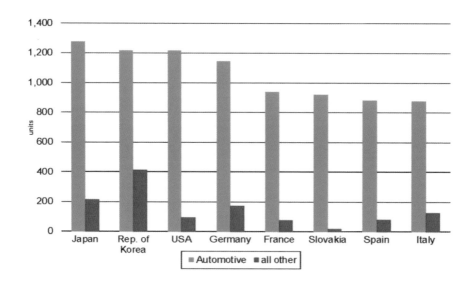

Source: International Federation of Robotics IFR [2015]

EXOSKELETONS

Related to the Human-Machine collaboration, it is worth noting the recent emergence of exoskeletons in the industrial sector. The exoskeletons are armor that serve to help operators to maintain prolonged postures and support weight in a more ergonomic way, which has applications in sectors such as automotive, aeronautics, naval, metalworking, logistics, energy and construction.

Today it may seem a far off technology, but these devices will live a very fast takeoff in the coming years. At least that is what the analysis firm Juniper Research predicts, which predicts a market of around 600 million dollars by the year 2022. It is about 900% more than the amount currently handled in this industry, around 53 million dollars.

Case study 36 : **SuitX** and its MAX modular exoskeleton (US).

SuitX, a spinoff of the University of California, develops the so-called modular **exoskeleton MAX**, consisting of three parts for areas of common injuries in the workplace: Shoulders, Lumbar and Knee. The launch cost is around $3000, varying depending on the units purchased.

Knees.- *LegX*, structure that goes from the hip to the feet with regulation of the degree of inclination and that allows the operator, with semi-flexed legs, to reduce the effort to stay standing (the sensation is similar to sitting on a chair)

Lower back.- With the *BackX* they manage to reduce the weight of the objects that are lifted from the ground by 13 kg.

Shoulders.- *ShoulderX*, which placed on the shoulders, allows you to reduce the effort when supporting weights on the head.

Source: SuitX

Case study 37 : **ExoArm** , low cost exoskeleton.

The exoskeleton options currently available have a generally high cost. But two young Slovenian engineers are developing a simple, open-source arm at a cost of only 100 €.

It's called **ExoArm**; it has an Arduino heart, and they're trying to program everything with an easy code to understand and modify. They have a first functional prototype capable of lifting weights of 10 kilograms, and the next step will be to improve the design to finally launch it on the market.

14. Artificial intelligence.

Probably there is no better way to describe artificial intelligence (AI) than with the definition of Marvin Minsky, one of the fathers of computer science and co-founder of the MIT AI laboratory: "It's the science of making machines do things that would require intelligence if people did them".

One of the most important objectives of AI systems is to reproduce human decision making but more quickly.

To have artificial intelligence we need Big Data, from the Internet of Things (IoT) and it is also necessary to apply algorithms that allow the machines to identify patterns of behavior, decision making, even anticipate our needs and learn autonomously. through reinforcement.

The **Machine Learning** concept or automatic machine learning has been the latest impulse to Artificial Intelligence. Using the right tools and processes, a machine can now learn better, faster and more reliably than a person starting from the same data.

Let's see the process broken down into 3 phases:

A. **Collection of process data**. We must collect as much data as possible from the process or machine. All these data, captured by sensors connected to a PLC or Controller, will be converted into information of interest.

B. **Transmission of data to the Cloud and Big Data**. After obtaining the data it is time to transmit them to a Cloud system to be treated and obtain information. This massive data collection process and storage is what we call Big Data.

C. **Algorithms and calculation models**. Once all the data is collected and stored, the next thing is to analyze them (**Data Mining**). This task is done through mathematical models and algorithms to reach conclusions and identify patterns and trends. This information will generate autonomous behavior of the machine and will make it learn based on the analysis of this data.

Currently there are quite a few data mining alternatives available. One of the best options is **Weka** which is developed by the *University of Waikato* (New Zealand), and is an extensive collection of algorithms that contains tools to perform tasks of classification, regression, clustering, association and visualization. It is freely available under the GNU general public license and is very portable because it is implemented in Java and can run on almost any platform.

Examples of Machine Learning Tools with open source:

✓ Apache Mahout

✓ R Stats Project

✓ Weka

✓ Vowpal Wabbit

✓ LibSVM , SVMLight

Sources: Sisteplant ; E. Dans ; B. Sanz; M. Guerrero; [2017]

Case study 38 : the **Smart Car** .

A clear example of Machine Learning in all its dimensions is found in the Smart Car, autonomous, capable of processing in real time a huge volume of data about road conditions, surrounding vehicles, signaling the temperature, environmental aspects and transforming everything into autonomous driving.

Case study 39 : Plastic **injection** process

Example of a company specialized in plastic injection of complex pieces of very thin wall:

The injectors allow the configuration by means of different parameters (inputs) of process (Melting temperature, nozzle temperature, injection time, injection pressure, Packing, Chilled time, etc.). The characteristics of Quality (outputs) are variables such as Hardness, Weight, Roughness, etc.

The current knowledge of the process is based on trial-error and past experiences, sometimes with paradigms that are not always true. Through **Machine Learning Tools**, the company has been able to dynamically visualize the effect of an input variable on one or more outputs (Simulation) and is capable of providing optimum operating values (Optimization)

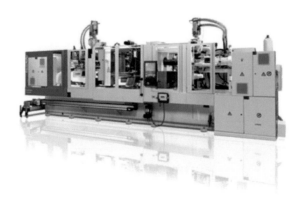

Case study 40 : Tool wear forecast.

A company has a dimensional control implemented in a machining process that performs various measurements that characterize the quality of the machined parts. The quality parameters get worse as the wear of the machining tool deteriorates. By means of Machine Learning tools they have been able to determine the time of substitution (also taking into account criteria of replacement cost, line performance, machine conditions, etc.).

In the initial phase the failure modes were analyzed (*Attrition flanks incidence, Crater wear, Wear by deformation, Sharp cutting, Fracture*) and the Symptom Analysis (*Roughness, Cutting forces, Consumed power, Temperature*), reaching the conclusion that <u>roughness</u> was the common factor.

In the next phase a 100% control of the degree of roughness of the machined parts was implemented in the line, which allowed them to obtain a real-time model of the process conditions and forecasting future roughness values, and therefore, the best moment for the useful change. *Source: Mitutoyo [2017]*

Case study 41: TI Automotive

TI Automotive, a leading global supplier of automotive fluid systems technology, has developed, in collaboration with Meifus and BCN Vision, a stamping machine that controls the dimensions of each unit produced by means of a complete artificial vision system.

All the parameters of the process and the values resulting from the inspection are stored with the same part code with which the production has been unitarily identified, so that future traceability can be guaranteed. In case the artificial vision detects that there is a drift of the required quality, the machine is able to self-regulate until reaching the optimum value of adjustment, thus avoiding the generation of defective parts in the process.

"The brain is not a glass to fill, but a lamp to light" Albert Einstein

15. Cloud Computing / Hosting

This is a technological model that allows companies and individuals to access a set of computer resources (both software and hardware) in a personalized manner and on demand through the Internet. Cloud storage makes all information accessible from anywhere. This facilitates access and analysis of the data throughout the entire value chain.

Cloud computing offers the possibility of using software solutions through the Internet, without the need for computer servers or experts who are responsible for the installation, maintenance and updating.

Advantages:

I. **Cost savings**. Payment for the use of products and services, eliminating additional costs such as the purchase of licenses, investment in IT infrastructure, equipment maintenance and updating.

II. **Storage and security**. There are providers that offer data storage services with virtually unlimited capacity. In addition, storage and backup services are included along with storage.

III. **Easy access**. Shared and real-time access to all information from anywhere and through any device with an Internet connection.

IV. **Easy drive**. System integration automatically, so companies do not need to worry about solving complex technical compatibility problems.

V. **Automatic updates**. The latest version of the software is always available.

VI. **Customization.** The systems in the cloud are customized according to the requirements and needs of the customer.

Source: AINIA Tech Center

16. Cybersecurity .

Concept that refers to the protection of information assets, through the treatment of threats that put at risk the information that is processed, stored and transported by the information systems that are interconnected. <u>The challenge of security is, without a doubt, one of the key points that companies must address.</u> The most important thing is the data and, even if an employee has the right device, if the information is not protected, corporate security is at risk. Today there are thousands of "doors" to protect and it is essential to have a strategy in the organization:

A. Elaborating rules and procedures for each service of the organization.
B. Defining the actions to be undertaken and choosing the people to contact in case of detecting a possible intrusion.
C. Sensitizing the operators with the problems related to the security of computer systems.

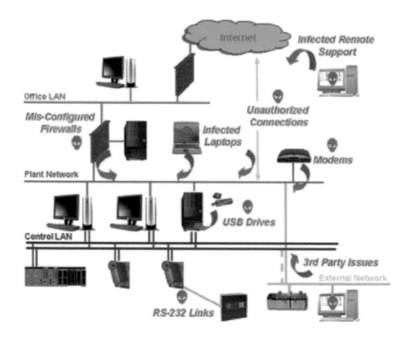

Pathways into the Control System Source: Control Engineering

17. Collaborative platforms.

A collaborative work platform is a virtual work space, that is, a computer tool that centralizes the functions of project management and knowledge, being accessible to all employees involved. The platform allows, therefore, the realization of specific work that can be shared with other users through the use of various tools integrated in the platform itself: a file exchange system, a forum, a chat, etc.

The companies that promote the use of collaborative platforms among their employees have the objective of the staff being proactive, committed and innovative and involved in strategic conversations that help to establish the future of the company.

The most important platforms are:

Yammer : This is Microsoft's proposal in this field. This platform, already implemented in a large number of companies, in its operation reminds us of a social network to use, with a timeline that is updated with comments, function of sharing updates of colleagues, personal profiles, aggregation of new contacts, groups of work and debate, internal messaging.

Trello: Trello allows us to create different boards in which to organize projects so that all team members can access and find their tasks and the status of each work organized in columns. In addition, each file not only brings together the subject to be dealt with, but can incorporate attachments, comments from the rest of the team, etc.

And others like **Producteev**; **SocialShared** and **Active Collab**.

Sources: Microsoft; TICBeat LAB; BSCW [2017]

Case study 42 : Philips

In 2013, Philips Spain launched the innovative digital platform "*Share Innovation*" to involve all its employees (companies, universities, research centers, etc.) in the proposal of technological solutions in all sectors. The collaborative digital platform is divided into four thematic areas that respond to current problems: *Habitable Cities, Health Trends, Personal Wellbeing and Responsible Innovation.*

Case study 43 : Telefonica

Telefonica, a global telecommunications company, uses internal social networks to promote the digital transformation of the company. In particular, they are using **Yammer** to debate and exchange knowledge and information in a fluid way, and thus drive innovation, problem solving and improve customer service.

Thanks to this collaborative platform Telefonica employees converse in 21 countries and exchange information in a productive, open, transparent way without hierarchies.

Real example:

• Following a strategic debate promoted by senior management, a group of "millennials" of the company voluntarily decided to create a community in Yammer and become "beta-testers" of new products designed for people who are part of that growing market.

Source: Microsoft [2016]

" *The future has many names: For the weak, it means the unattainable. For the fearful, it means the unknown. For the courageous, it means opportunity. " Victor Hugo.*

18. PLM (Product Lifecycle Management)

PLM is a new and ambitious business computing solution that aims to manage the entire life cycle of the product on a single computing platform: from the time it is created and designed, through the prototype phase, first pre-series, manufacturing and maintenance-after-sales service.

A PLM system coordinates how people create and use product information in their daily processes. PLM systems integrate the islands of information existing in companies, caused by sequential, fragmented processes, based on scattered papers and files and with a lot of manual intervention. Without PLM, new product launches are slow, consumers of resources that are scarce, have little visibility, and are difficult to manage and control.

And when a product is intelligent and connected, the data generated during the entire life cycle no longer ends when the product is shipped. New *Internet of Things* technologies allow companies to capture and use information about the performance of a product during its operation, with the potential to improve future products.

Benefits in the execution of the business:

a) Reduce costs thanks to better access to consistent data.

b) Increase business opportunities.

Benefits for the organization:

a) Eliminates geographical barriers and facilitates internationalization.

b) Helps make changes in the organization.

c) Benefits for the product or service

d) Encourages the reuse of standard components and previous designs.

e) Facilitates the definition and modular management of the product

The main functions of a PLM system are

a) Store, organize and protect the data

b) Manage documents and their changes

c) Search and retrieve information

d) Share data with users in a controlled manner.

e) Execute processes and workflows.

f) Display data and documents

Sometimes the terms PLM and PDM are used interchangeably but there are differences:

PDM (Product Data Management): Refers to the management of design files and lists of parts and specifications. Basically, it is about organizing the files and creating identification and revision rules for each item, creating a digital library of CAD files and other types of design files.

Sources: EOI; Siemens [2017]

Currently the most interesting applications are:

Dassault (**ENOVIA**), **Siemens** and **SMARTEAM**.

Example ENOVIA application

Example SMARTEAM® application

PLM is the ideal tool to create a Digital Twin, as we will see below.

"The value of innovation is not to prevent others from copying you, but to get everyone to try to copy you"
Enrique Dans

19. Digital Twin (Cyber-Physical Equivalence CPE)

The concept of a *digital twin* (coined by NASA) works exactly as it sounds. It refers to computerized duplication of physical assets that can be used for various purposes. They use data from sensors installed on physical objects to represent their status, working conditions or position in real time. A digital twin is, therefore, a software model of an installation or machine that uses sensor data to understand its status, change as it changes, improve its operability and add value.

In the same way that we talk about the digital twin in the development of the product, it is also feasible to use a virtual model to define the manufacturing processes and their subsequent simulation. We can do it from a level of operation to the level of production line, even reaching the simulation of the entire plant. You can test changes before you start manufacturing.

<u>Only when all requirements are met virtually, is it physically produced.</u>

The digital twin also facilitates decision making in an objective way in terms of investments in means of production and allows us to achieve a maximum optimization of available resources.

Case study 44: Siemens and Land Rover BAR

Siemens has created a digital twin for Land Rover BAR, the British team of the America's Sailing Cup.

During training and races, hundreds of ship's channels and crew data are transmitted directly to the ground. Sensors and video technology communicate in a very precise way the performance of the boats so that the designers and the engineers analyze it in the **Mission Control**. The digital twin responds exactly like the real one to any test to which they submit it virtually. The tools that Siemens has contributed to the race are common in automotive or aerospace technology.

Sources : Siemens; Land Rover BAR; Robin Hancock [2017]

" A person with a new idea is a crank until the idea succeeds "

Mark Twain

Case study 45 : Digital Twin in **Wind Farms**

General Electric has implemented many low cost sensors in wind farm equipment in order to analyze all the information.

"It is expected that these turbines operate over a period of several decades in some of the harshest environments in which a machine can be subjected. If they are going to fail, we want to know as soon as possible. A wind farm is a very unreliable source of energy. It is not like the nuclear, gas or coal where you know the energy output you're going to get. With the digital wind farm, we are helping to make daily forecasts, react and deal better with high vertical integrated markets such as North America and Europe" says *Sham Chotai*, Technology Director at General Electric.

The company estimates that it could increase a park's energy production by 20%.

Sources : Sham Chotai , GE Power & Water [2017]

"The best way to predict the future is to invent it."
Alan Kay

20. Cyber-physical systems (CPS)

CPS implies a multidisciplinary approach, merging various KETs that we have seen in previous chapters. The most complete definition is provided by ECSEL, *"Generation of integrated, intelligent, interconnected, interdependent, collaborative and autonomous ICT systems that provide computing and communication, as well as monitoring and / or control of physical components / processes in different application domains"* .

The Cyber physical Systems integrate the <u>3 fundamental pillars of Industry 4.0:</u>

1) *Horizontal integration of the information systems* that are used in the different links of the supply chain, and that until now were disconnected silos of information.

2) *Digital integration of engineering processes* (PDL Product Lifecycle Management).

3) Vertical integration of information systems (sensors, control systems, SCADA systems, Digital Twin, ...)

To create a complete cyber physical space we will previously need Digital Twins. By having a real physical space connected to the digital, we can interact without limits, for example, from preventing an engine from running until the engine is turned off.

When the term CPS is applied to an exclusive production context, it is called **CPPS**, or cyber-physical production systems.

Case study 46 : Smart Grid.

The smart grid is a form of efficient electricity management that uses a CPS system to optimize the production and distribution of electricity in order to better balance the supply and demand between producers and consumers, allowing the coordination areas of protections, control, instrumentation, measurement, quality and energy management, etc., that are linked in a single management system.

The irruption of renewable energies in the energy landscape has changed the energy flows in the electricity grid: now users not only consume, but also produce electricity through the same network. Therefore, the <u>flow of energy is now bidirectional</u>, and management is more complex.

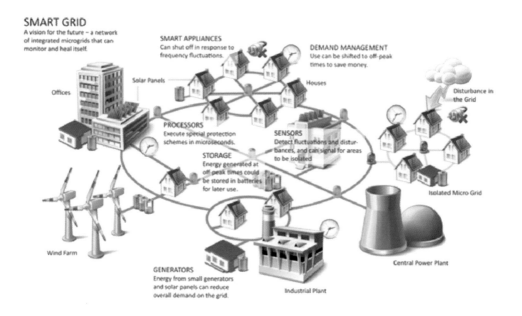

Source: Meinberg [2017]

Case study 47 : Automatic Aeronautical Pilot.

The aircraft autopilot is another clear example of CPS systems. The autopilot is integrated in what is known as a flight management system, known by its acronym in English FMS.

The FMS coordinates the different electronic systems of the airplane through which it tries to have controlled the flight conditions and also allows us to interact with instrumental landing systems (ILS).

Commercial aircraft manufacturers are already studying the introduction of unmanned aircraft (*UAV or unmanned aerial vehicle*), in principle for the transport of goods, and in the future, including passenger. The planes would fly with a combination of an autopilot and a remote (human) pilot on the ground.

"Success consists of going from failure to failure without loss of enthusiasm."
Winston Churchill

Case study 48 : Nuclear Power Plants, GE CPS.

General Electric works on the development of CPS for power and nuclear power plants, based on the *Predix* platform.

"In the case of a nuclear plant, if a compressor fails, it can cause a shutdown and cost millions of dollars to restart the plant again. Using the concept of the digital twin combined with the deep learning of the machines, we are able to predict 30 to 60 days before a compressor will fail" says Sham Chotai, Technology Director at General Electric.

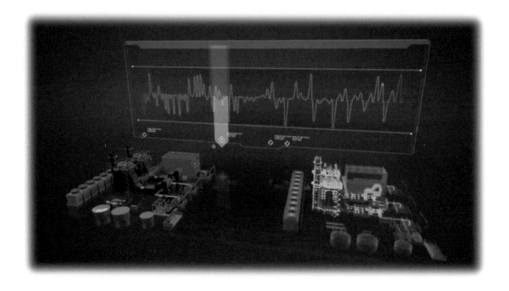

Sources : Sham Chotai , GE Power & Water. [2017]

"Man's mind, once stretched by a new idea, never regains its original dimensions."
Oliver Wendell Holmes .

5. MATURITY INDEX

MODEL OF SELF-EVALUATION IN THE EIGHT KEY AREAS OF VALUE GENERATION IN THE INDUSTRY 4.0.
Source: Own elaboration

Each sub section appears valued between 1 and 5 points, we must identify the answer that best fits our situation today and the total sum will give us **maturity level** *of our company in relation to the Industry 4.0*

		Score		Result	%
		Minimum	Maximum		
1	OPERATIONAL PROCESSES	3	15		
2	INDUSTRIAL ASSETS	3	15		
3	ENERGY	3	15		
4	PEOPLE	3	15		
5	QUALITY	3	15		
6	SUPPLY-DEMAND SYNCHRONIZING	3	15		
7	TIME TO MARKET	3	15		
8	INTERNAL LOGISTICS AND SUPPLY CHAIN	3	15		

1. OPERATIONAL PROCESSES

Sensoring , monitoring and control

1. Standard sensorization
2. Sensing and measuring multiple process parameters
3. Descriptive analysis and visualization parameters
4. Advanced analytics (predictive and prescriptive analysis)
5. Autonomous control of operational processes

Intelligent processes

6. Manual processes predominate over automated
7. High degree of automation
8. Collaborative robotics applied to business processes
9. Processes with Artificial Intelligence and Machine Learning
10. Fully intelligent and autonomous processes

Virtualization

11. Independent operational processes management software

12. Implementation of monitoring, control and data acquisition systems (SCADA)

13. Implementation of MES and CMMS systems

14. Partial virtualization with application of RA and RV systems in the operational processes

15. Complete virtualization through cyber-physical systems (CPS)

2. INDUSTRIAL ASSETS

Flexible manufacturing and modular systems

16. Batch manufacturing system and static layout

17. Quick reference changes (manuals); processes, machines and people with high versatility and flexibility

18. Automated reference changes

19. Additive manufacturing incorporated into the manufacturing process

20. Modularity: Total Flexibility to replace, add or remove new elements

Access and Remote Control

21. Access and control of industrial assets physically in the plant

22. Remote access to the status / information of industrial assets

23. Remote control of industrial assets

24. Autonomous control through AI and Machine Learning

25. Access and control of all assets through the CPS with a high degree of cybersecurity

Predictive Maintenance

26. Corrective maintenance is dominant

27. Higher percentage of preventive maintenance

28. Meaningful maintenance management with predictive systems

29. Predictive maintenance with digital twin application in key machines or processes

30. Predictive maintenance fully implemented and controlled by a cyberphysical system

3.   ENERGY

Monitoring and Control

31. Information consumption through energy trading companies

32. Entering data logging sensors for control of energy consumption

33. Monitoring consumption in real time

34. Compared with consumption patterns and alarm generation

35. Systems that identify inefficiencies and propose corrective actions

Smart consumer

36. Conventional power management

37. Energy audits

38. Implementation of advanced energy saving systems

39. Energy factor incorporated into the products / processes design phase

40. Green Technology (Green IT) fully implemented

Efficient energy systems

41. Energy consumption on demand

42. Control of energy demand

43. Power self-generation (secure, sustainable and competitive)

44. Energy storage systems implemented and well-balanced energy demand curve

45. Company without dependence on the energy network which even generates power with a positive net balance

4. PEOPLE

Digital training

46. Basic knowledge of new technology

47. Employees and partners know the technology that makes possible Industry 4.0, and the changes and benefits of its application

48. The new skills required have been identified and there are some guidelines to follow for activation of people talent

49. *Digital Enablers* available inside and outside the company

50. Employees and partners with full digital training

Interfaces

51. Person-machine communication in-situ

52. Remote control devices

53. Abandonment of routine tasks. People focused on creative activities with added value

54. Ubiquitous access to all devices through friendly interfaces

55. Independent monitoring , interacting with all devices

Human-Cyber-Physical Systems

56. Traditional system of collaboration / communication between people and systems

57. Horizontal integration of information systems (supply, manufacturing, sales, etc.)

58. Digital integration of engineering processes (PLM)

59. Vertical integration of information systems (digital twin)

60. Fully integrated cyberphysical system, monitoring and controlling physical processes in an intelligent and autonomous way.

5.   INTERNAL LOGISTICS AND SUPPLY CHAIN

Warehouse management

61. Without automating.

62. Semi-automatic warehouse: vertical cabinets, paternoster

63. Warehouse and automatic picking: air transporters, AGVs

64. Automatic warehouses integrated in the chain.

65. No warehouses in the supply chain (complete synchronization with only consolidation points).

Internal logistics

66. Push Supply process .

67. Pull Supply process.

68. *JIT* delivery.

69. Synchronized delivery with the sequence of production.

70. *Full kitting* deliveries.

Manufacturing supply

71. Manual supply.

72. Manual logistics train.

73. AGV with fixed routes.

74. AGV with open routes.

75. AGV led by Manufacturing Department

6. QUALITY

Unitary Quality Control

76. Frequency quality control

77. Quality control unit Unitary quality control

78. 100% quality control of all parameters with advanced control systems (artificial vision)

79. Machine Learning systems and automatic adjustment of machine parameters

80. Non-Quality Predictive Systems

Digital Quality Management

81. Traditional Quality Management

82. Automation and digitalization of Quality controls

83. Using Big Data and Smart Data for Quality Management

84. Integrated quality as a key axis in PLM platform

85. Total digital quality management through cyber-physical systems

Full traceability in the value chain

86. Batch traceability (with product parameters)

87. Batch traceability (with parameters and process conditions)

88. Unit traceability in the production process

89. Unit traceability in the supply chain

90. Full traceability in the value chain

7. SUPPLY-DEMAND SYNCHRONIZING

Product tailored to customer based on data

91. Traditional design based on experience and / or intuition

92. The company has launched some of the 4 intelligent industry trends: *(1) Use of internal (employees) and external collaborative methods (customers, universities, research centers); (2) Using powerful communication channels to reach prospective client and make him share their product design; (3) Predictive Analysis of customer needs through Big Data systems; (4) massive design customized products.*

93. The company has reached technological maturity in 2 of them

94. The company has reached technological maturity in 3 of them

95. The company has reached technological maturity in all trends

Customer logistics

96. Push Delivery Process

97. Shipment management based on actual orders

98. Active management of deliveries (high flexibility and reduced lead time).

99. Automatic delivery management (communication between own and customers systems to optimize deliveries).

100. Predictive demand systems

Logistic routes

101. Decentralized vehicle fleet (contracting with several suppliers).

102. Centralized fleet vehicles.

103. Centralized and planned fleet (optimized routes)

104. Route and navigation in real time (routes are configured and fed back in real time to increase optimization

105. Autonomous Transportation.

8.   Time to Market

(Ability to bring to market new products and get feedback from the customer)

Innovation Process

106. Classical innovation process ("funnel of ideas" model)

107. Technology push innovation model (sequential and orderly process based on scientific knowledge that ends in an economically viable product)

108. Market pull model (the needs of consumers activate the innovation process)

109. Open innovation (incorporation of knowledge and technology from within the organization abroad)

110. Open innovation and co-creation at all levels (suppliers, customers, partners and Community)

Product life cycle

111. Traditional Life Cycle Management Process

112. File management with PDM (Product Data Management) system

113. PLM solution used in some of the phases (design, manufacturing and after-sales)

114. PLM solution fully implemented

115. Digital Twin used for the development of the product, the definition of the production processes and their subsequent simulation (virtual engineering)

Distribution and Sales

116. Conventional distribution and sales system

117. The company has initiated some of the 4 smart industry trends:
(1) Real-time connection between company / product and client: from before your purchase to the shipping process, the final reception and the after-sales;
(2) Scanning of channels: the customer can buy in 24/7, accessing from any device and place;
(3) Effective and personalized distribution: higher and more agile quantity of shipments;
(4) Prediction of consumption habits

118. The company has reached technological maturity in 2 of them

119. The company has reached technological maturity in 3 of them

120. The company has reached technological maturity in all trends

"There is no favorable wind for those

who do not know where they are going"

Seneca .

6. CURRENT TECHNOLOGY EXPECTATION

The current technological expectation is constantly being analyzed by Gartner in the **Hype Cycle** graphic, and today is considered the benchmark to follow.

The Hype Cycle for Emerging Technologies is a representation of the maturity, adoption and application of specific technologies. The report Hype Cycle is published on a yearly basis and provides a cross - sectional view of trends in emerging technology industries, helping to discern if we are faced with excessive expectation or a viable technology.

Each Hype Cycle is comprised of five key phases of the life cycle of a technology, launch, peak oversized expectations, the abyss of disappointment, the ramp consolidation and finally the plateau of productivity.

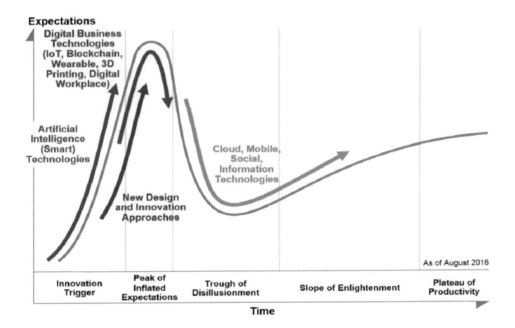

Launch (Innovation Trigger):

The beginning of the cycle appears next to a potential technological advance and the appearance of the first news about the new concept. The interest of the media causes a significant appearance of publicity around the new technological development. Often there are still no products that apply the technology and commercial viability is not proven.

Oversized expectations peak (Peak of Inflated Expectations):
Early advertising produces a series of success stories, often accompanied by dozens of failures. Some companies take measures in relation to the new technology, although many do not and they wait to see its evolution.

Abyss of disillusion (Trough of Disillusionment):
Interest decreases when there are new experiments and implementations that use the technology. This phase depends on the distribution by the producers of the new technology, and continues to invest only if suppliers that are committed to technology can satisfy the early adopters through product improvements.

Consolidation ramp (Slope of Enlightenment):
Examples of how technology can benefit businesses begin to crystallize and become more widely understood. Technology vendors launch products of second and third generation. A larger number of companies invest in technology while more conservative companies remain cautious.

Plateau of Productivity:

The widespread adoption begins to take off. Criteria to assess the viability of suppliers are defined more clearly. The broad applicability and relevance of technology in the market are paying off.

WHAT USE IS THE HYPE CYCLE?

The Hype Cycle offers a vision of how a technology will evolve over time, providing insight into the penetration of new technology that allows managing future impact in the context of business objectives.

Professionals from different sectors use the Hype Cycles to obtain information about the evolution of emerging technologies in the context of their industry, and to act according to the risk they pose. In this context Gartner distinguishes three approaches depending on the risk posed by different times of adoption of a technology.

The early adoption of an emerging technology combines risk-taking, because although it may receive the benefits of early adoption, the technology may not mature and may not yield the expected result.

Depending on the position of the emerging technology within the cycle, it may be more advisable to act moderately assuming the cost / benefit ratio of implementing a new technology that has not yet been fully tested, waiting until others have been able to offer tangible value.

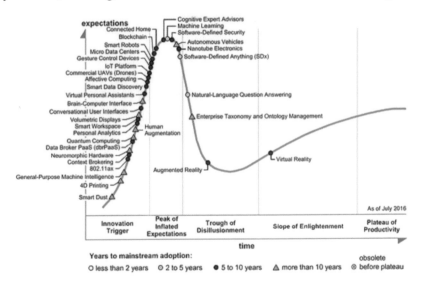

Source: <u>Gartner Research Methodologies</u> (July 2016)

"I cannot understand why people are scared with new ideas. I am scared with the old ones". John Cage

7. THE ROLE OF LEAN IN THE FACTORY OF THE FUTURE

Lean Manufacturing (or simply Lean) is a work philosophy that seeks to improve and optimize production systems by eliminating "waste". In this case, by "waste" we mean the processes that use more resources than necessary: Lean will not disappear with Industry 4.0, rather on the contrary; Lean principles will probably become more important. With the 4th industrial revolution the real Lean Company may develop.

Industry 4.0 allows a better understanding of customer demand and allows immediate sharing of information, therefore we can produce faster, with less waste and produce exactly what the customer needs: Lean objective: avoid overproduction, losses of time, excess inventory and unnecessary movements.

Industry 4.0 converts manual tasks, monotonous and repetitive, into supervision and control tasks: Lean objective: to avoid underutilized human potential.

Industry 4.0 drives manufacturing models with intelligent machines and equipped with machine learning or self-learning, which are regulated and adjusted autonomously to manufacture 100% compliant products: Lean objective: avoid waste and scrap.

To summarize, the technologies of Industry 4.0 may be exactly what we need to consolidate lean manufacturing at all levels of our organization.

"If you want different results, do not do the same". Albert Einstein.

ABOUT THE AUTHOR

Francisco Yáñez Brea, Spanish engineer and Plant Manager of *TI Automotive* (Leading Manufacturer of Automobile Components), which has recently been recognized by the French carmaker *PSA Group* (which includes brands Peugeot, Citroen DS, Opel and Vauxhall) as one of its best suppliers (*best Plant 2017 Award*).

As a member and partner of CEAGA (first automotive cluster in Europe with the recognition " Gold Label ") he has participated in the development of the *Road Map 4.0* ,a strategic document prepared for the purpose of laying the foundations for a technology roadmap with which to drive the adoption of Industry 4.0 in companies in the automotive sector in Galicia.

Printed in Great Britain
by Amazon